Water (H2O) Settles Proton Outward

Understanding Amazing Chemical Properties of Water

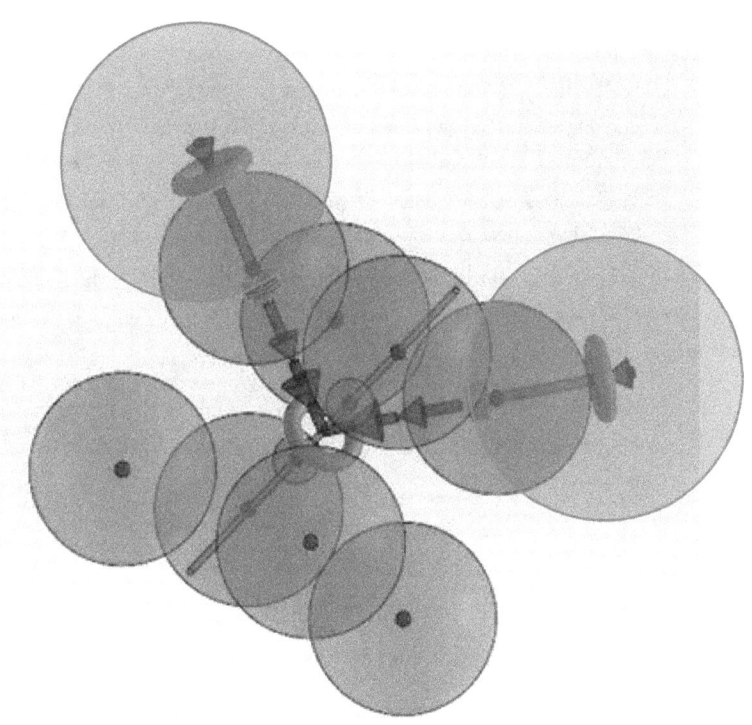

By Arno Vigen

© 2018 E Arno Vigen

Table of Contents

Electrons Have Knowable Settling Positions 3
    The Oxygen in Water Wants a Full Shell 4
    Scrunching – Why Oxygen Bonding Angle 104.5 Degrees versus perfect Tetrahedron 109.5 of Carbon 5
        Note that Working Receivers Usually Do Not Extend Beyond Outer Electron Repulsion Zone. 11
Hydrogen has One Proton and One Electron, and Electron Fills the Oxygen Open Position 14
Water and the Ideal Gas Law 18
Water and pH 23
    Sizing Hydrogen Proton Attraction Zones as Larger than Electron Repulsion Zones 23
    Percentage Reactivity of Hydrogen (pH) 24
    Low pH Means Water has Extra Hydrogen on Every Free Oxygen Electron 25
    The Thylakoid CO2 Breaking Photosynthesis Process 27
    Standard Hydrogen Atom does not Break Bond with CO2 29
Hydrogen Bonds and Water 33
In AVSC Chemical Engineering Tool, Electrons (and Protons) have Repulsion Zones 35
    When Atoms Meet, the Repulsion Zones Operate Like Rubber Balls 38
    Understanding Bonding Angles for Carbon (109 degrees), Nitrogen (107), and Oxygen (104) 40
    Hydrogen Drives Exceptions to the Closed Space Ideal Gas Law 41
Conclusion 43
Endnotes 47

Electrons Have Knowable Settling Positions

Current universities teach that electrons cannot have known position and speed. They only operate in a statistical cloud.

i

However, that is not the case. Yes, there are no tiny iron frameworks holding electrons in their positions, and yes that means that electrons can wobble around in harmonics with the surrounding electrons. Instead, particles have relative positions as a set.[ii] However, subatomic particles have a settling position based upon the total forces around them. Atoms work as a group with electrons in natural positions based upon the surrounding set of subatomic particles.

One of the most amazing results of this revised view is that we can determine all the amazing qualities of water (H_2O).

The Oxygen in Water Wants a Full Shell

The Oxygen in water has six (6) electrons in its outer shell, but it wants eight (8); eight being a full tetrahedron in each hemisphere of its magnetic field (2 x 4 = 8). That is natural level for Shell-2 and Shell-3 in the Periodic Table.

Two tetrahedrons are a cube. You can see this in the below; the yellow tetrahedron goes to every other corner of the cube. That leaves four (4) green corners which then forms a matching tetrahedron.

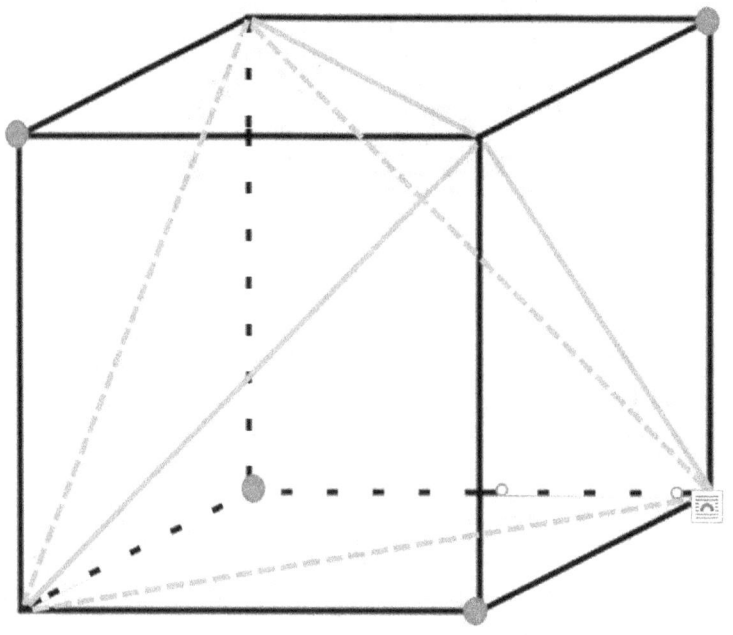

A slight nuance is that the tetrahedrons are anchor at the nucleomagnetics poles. The two electrons settle at the magnetic poles. It is well known that two electrons always find those polar (closer) positions. Those are known as 2s1,2s2 in traditional naming, and known as 2m1, 2m2 in AVSC[iii] since m=magnetic pole for easier remembering.

Scrunching – Why Oxygen Bonding Angle 104.5 Degrees versus perfect Tetrahedron 109.5 of Carbon

Nitrogen and Oxygen have this special quality where the extra electrons above the one tetrahedron (4) squeeze or scrunch the cube. The two anchors are at opposite corners, and they want to settle at closer positions, so the tetrahedron overlap, and the outer electrons move off 70.5 degrees nucleomagnetics axis angle to 66 degrees for Oxygen. That creates an angle between them, the bonding angle of 104.5 degrees.

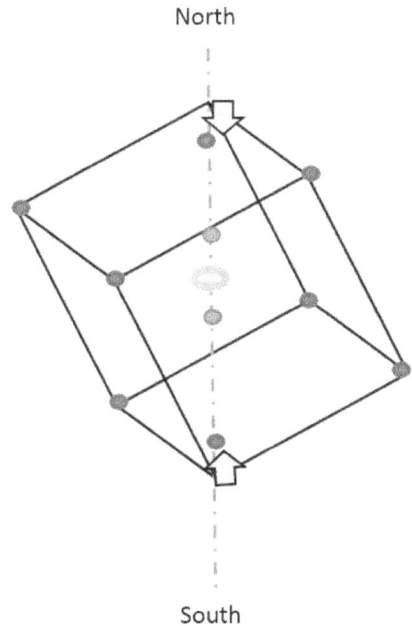

That means that 08-O Oxygen has eight (8) electrons, but two are in the Shell-1 on the nucleomagnetics axis very tight to the nucleus. Oxygen has six outer shell-2 electrons. Four (4) are one tetrahedron, which leaves the second tetrahedron with one on the nucleomagnetics axis filled, and one electron with two open positions for bonding as settling positions of known inclination/longitude and latitude.

Electron	Inclination / Longitude	Latitude / Azimuth
Shell-1:		
e-1m1 (1s1)	0 at the pole	n/a
e-1m2 (1s2)	180 at other pole	n/a
Shell-2 – Tetrahedron-1:		
e-2m1 (2s1)	0 at the pole	n/a
e-2c1 (2p1)	66 degrees	0
e-2c2 (2p2)	66 degrees	120
e-2c3 (2p3)	66 degrees	240
Shell-2 – Tetrahedron-2:		
1m2 (1s2)	180 at other pole	n/a
e-2c4 (2p4)	66 degrees	60
b-2c5 (2p5)	66 degrees	180
b-2c6 (2p6)	66 degrees	240

Now, the electron may wiggle around, but the electrostatic attraction to nucleus protons brings them back to these positions. The electron-electron electrostatic repulsion keeps them in positions. Finally, the nucleomagnetics force pushes them outward into that funny magnetic shaped field; that is what causes the first two electron to go to the poles (Hydrogen, then Helium, and so on). Oxygen having more count in the nucleus means that it has a stronger, wider magnetic field which pushes electrons into those particular angles relative to the nucleomagnetics axis.

The yellow line where a Carbon nucleomagnetics field puts the 06-C Carbon black electrons is different, weaker, and thereby closer. This compares to the black line where the Nitrogen purple electrons would settle. Further, the field would grow more so the red 08-O Oxygen electrons are further 'scrunched' outward. (Distances exaggerated for visual effect.)

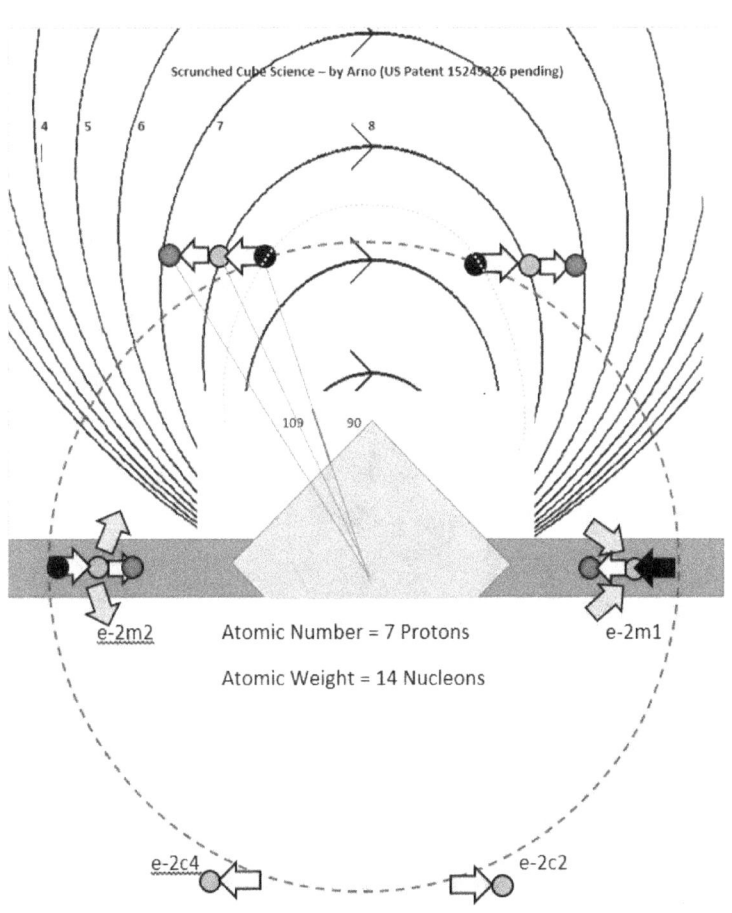

So, 06-Oxygen has two bonding positions as specific angles. It has two open positions at the know angle of 104.5 degrees.

The Receiving Bonding Positions (inward arrow) have specific angles based upon the orientation.

Equator View

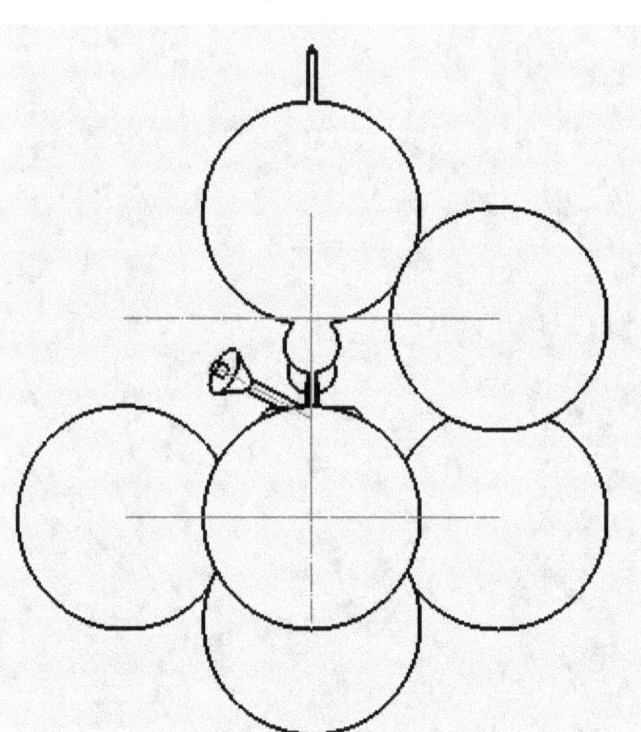

From the 2m1 polar view, you cannot see the receiving bonders as they are in the other hemisphere.

Polar 2m1 View

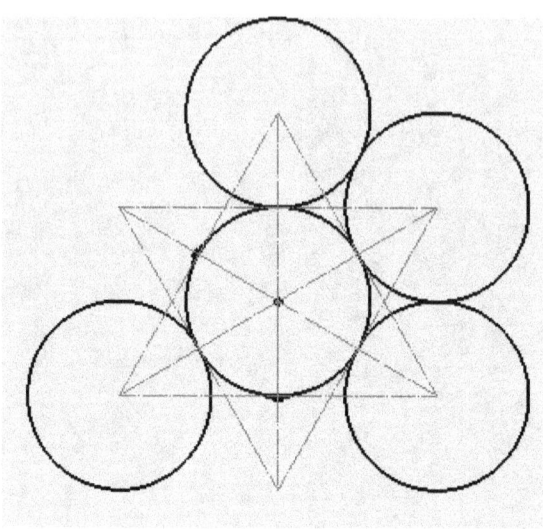

Isometric View

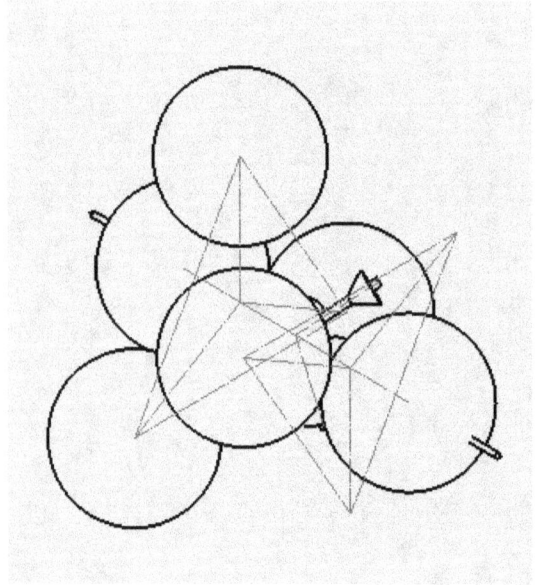

Note that Working Receivers Usually Do Not Extend Beyond Outer Electron Repulsion Zone.

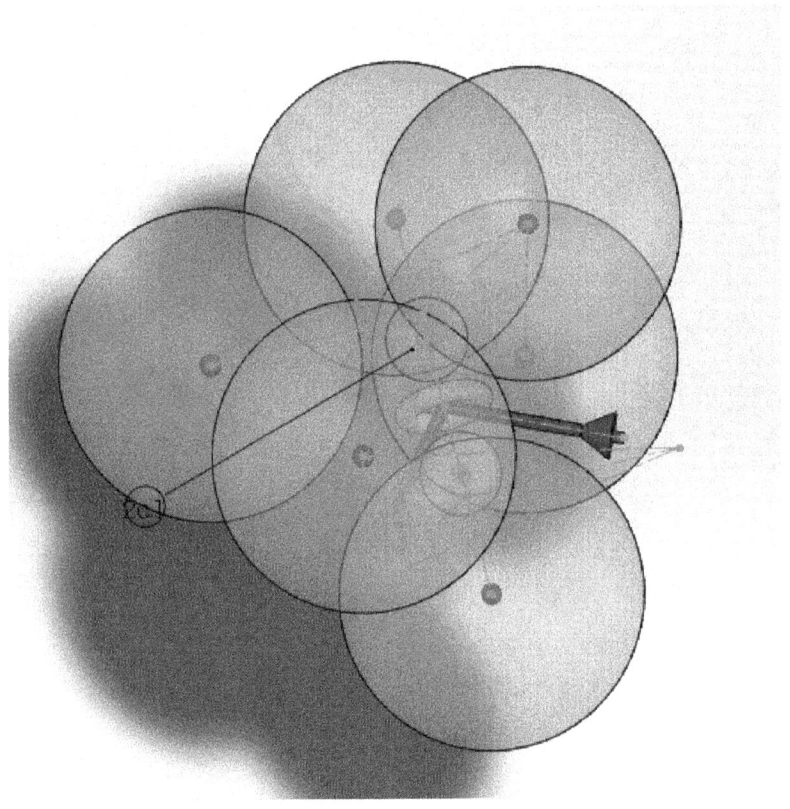

H2O Equator View

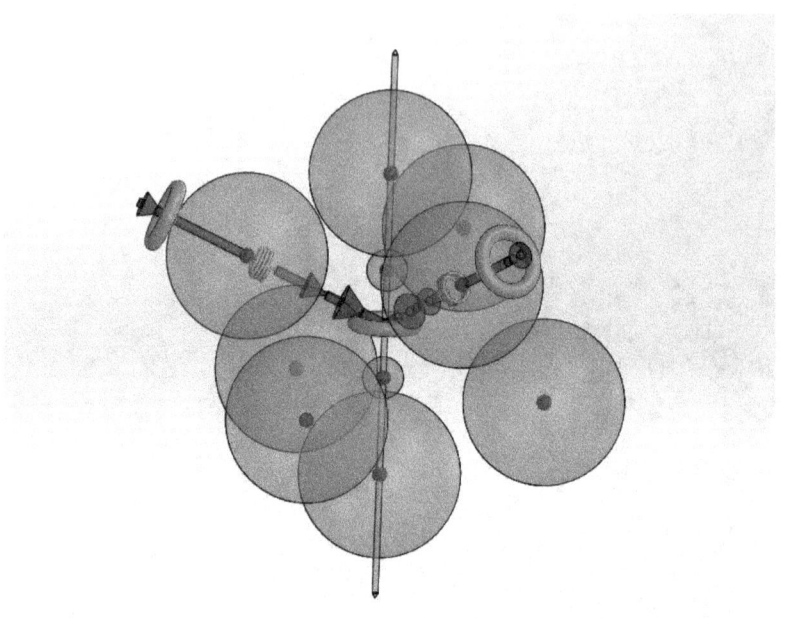

That creates a structure, relative to the nucleus and its nucleomagnetics axis, for each atom which is stable. That is, the electrons move generally in a 'valley' at specific inclination / longitudinal angles to that axis, and latitude / azimuth angles relative to each other.

Equator View: Outer Subshell Structure
07-O Oxygen

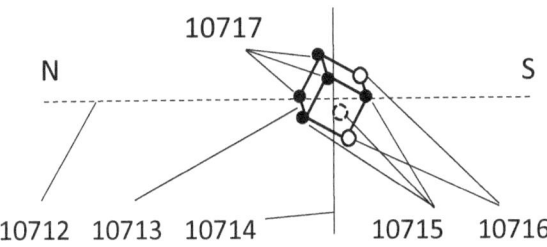

Hydrogen has One Proton and One Electron, and Electron Fills the Oxygen Open Position

A Hydrogen has one electron (blue sphere) and one Proton (gray ring). Hydrogen wants to contribute (red outward arrow) in the electron direction. It wants to receive (blue inward arrow) in the nucleus proton direction.

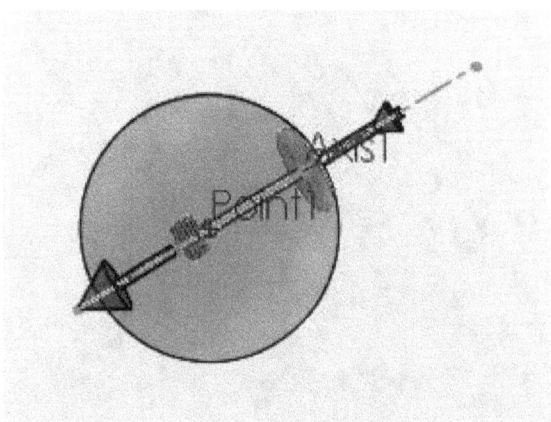

That is part of the magic of Hydrogen. It can be either a contributor and a receiver. In fact, in explosive hydrocarbons or amino acids different Hydrogens in the same molecules express some as contributing and some as receiving.

That said, the electron in Hydrogen for H2O wants to settle into the open Oxygen position. It 'sees' and open path of electrostatic attraction. That open view is a place of attraction (electron-proton) versus repulsion (electron-electron).

In AVSC engineering tools, each electron has a bluish repulsion zone which makes it easy to visualize the location of open bonding positions. To bond, the engineer must have a 3D view towards the nucleus (ring).

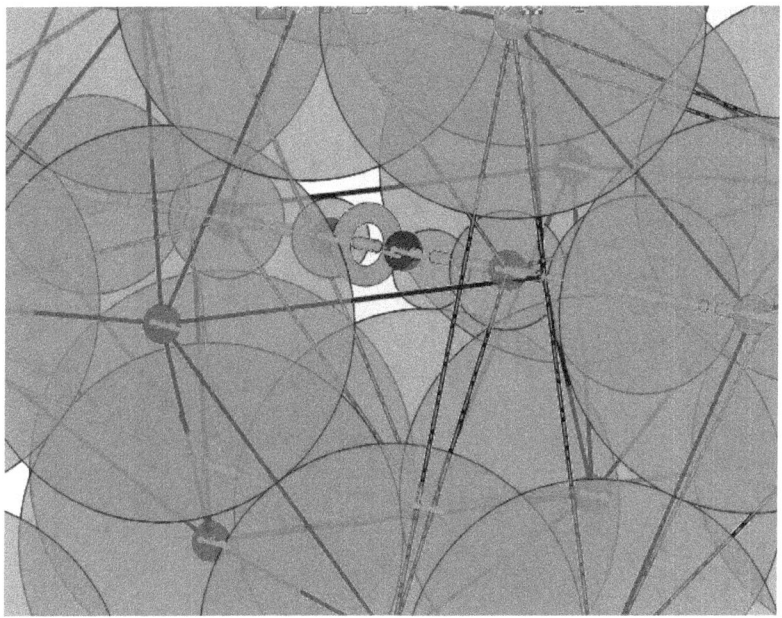

That process creates a H2O that is different. The electron settles in a position in the Oxygen shell. One electron (not a Pauli pair) creates the bond. (The Pauli paired electrons are actually on the far size of the Oxygen in the opposite 'spin' hemisphere. In an atom, if one Pauli electron moves right, the whole structure rotates, so the opposite hemisphere 'spin' electron moves left. Hence, we have opposite 'spin'. However, there are not two electrons in a bond. Only one electron fills an outer shell position in both the atoms so bonded).

In the final engineering drawing by AVSC of water (H2O) you have the nucleomagnetics axis of the 08-O Oxygen (a yellow axis). That Oxygen has a full outer shell with eight (8) electrons, but two of those are also attached to Hydrogen nucleus (proton). As such, those the proton nucleus of the two 01-H Hydrogen are facing outwards.

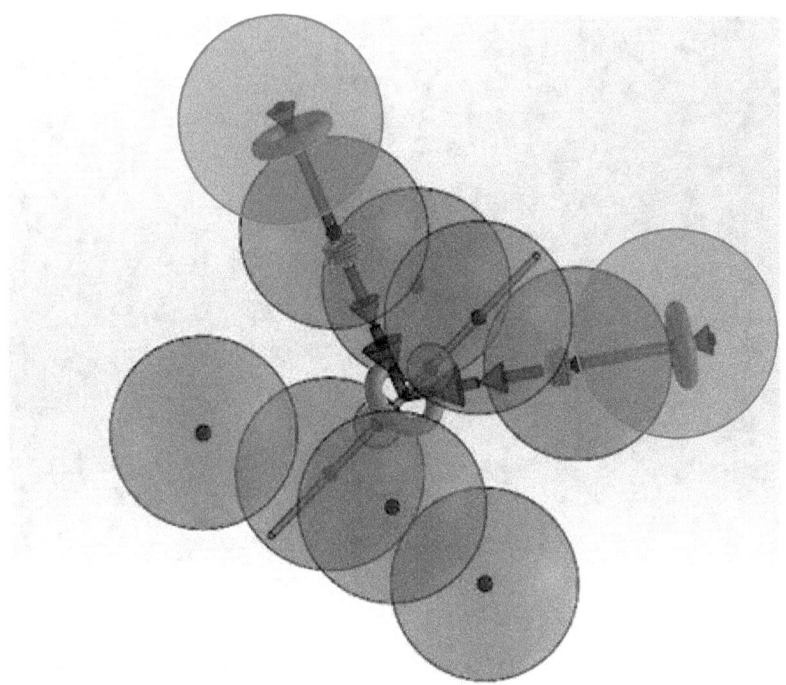

In AVSC Engineering Tools, we will further show those positions with Protons outward with a reddish Attraction Zone. That compares to the bluish repulsion zones of the standard atom's electron outer settling positions.

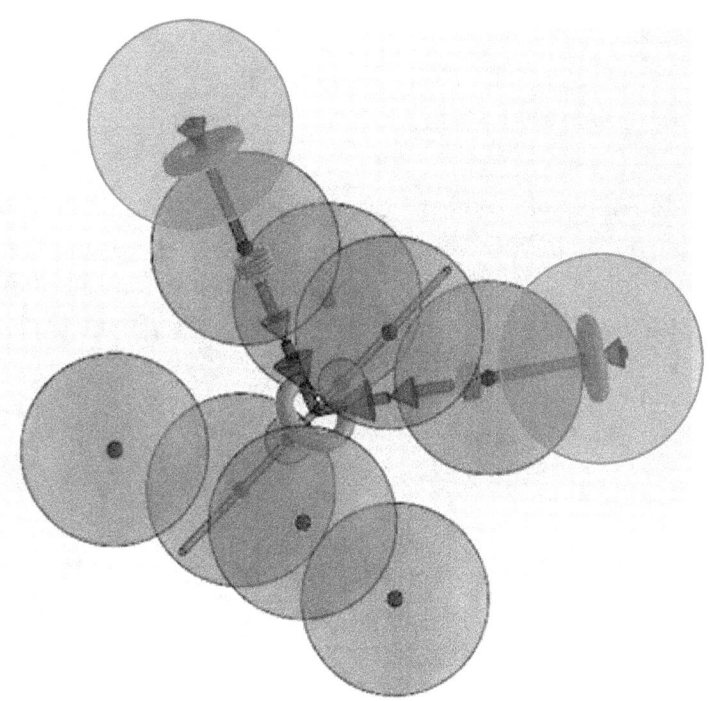

Water and the Ideal Gas Law

There is a well know formula for the separate of atoms when in the gas state. It is known as the Closed Container Ideal Gas Law:

$$PV=nRT$$

Atoms when build by AVSC engineering have lots of electrons on the outside. However, there are not perfectly spaced. There are directions where the repulsion zone is weaker, and another atom's outer electron might form a temporary bond (liquid) or permanent bond (solid). Remember the picture.

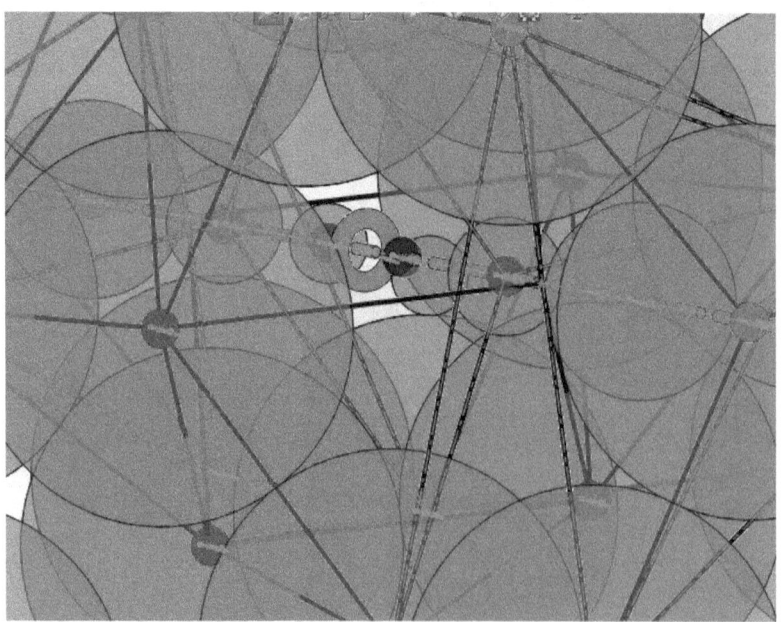

While an electron might find one of these spaces, this creates a long torque from one molecule to another. That makes a very small force able to break that bond. The base atom outer electron acts like a fulcrum. The other atom outer electron then can easily 'break' off. It might bond, but that bond is very weak based upon the ratio of the angle to its own nearest electron versus the full length to the other atom's setting position.

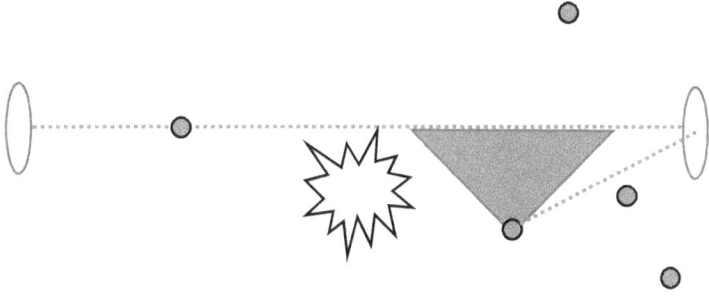

That 2nd atom's electron settling-position in temporary, liquid bonds is usually about 3-4x the average radius of base atom. That is big fulcrum. It easily breaks off.

TRY THIS: Hold your arm out. Put a reasonably heavy object on your inner elbow. Easy enough. Now put that same in your hand. That is twice (2x) the distance, but four (4x) the force (2^2). Now imagine holding a yardstick (3 more feet) with the same object another arm's length away. That is 16x the force (4^2). You can imagine that the stick will easily break under the same weight.

However, for solid bonds, like Hydrogen into Oxygen for H2O, that bonding distance is not far out. In fact, it is nearly the same as the traditional bonding distance. (The Hydrogen proton is pulled in by the other Oxygen outer electrons. This is the physical 3D engineering model of all bonding in AVSC.)

H2O 2D Settling Positions

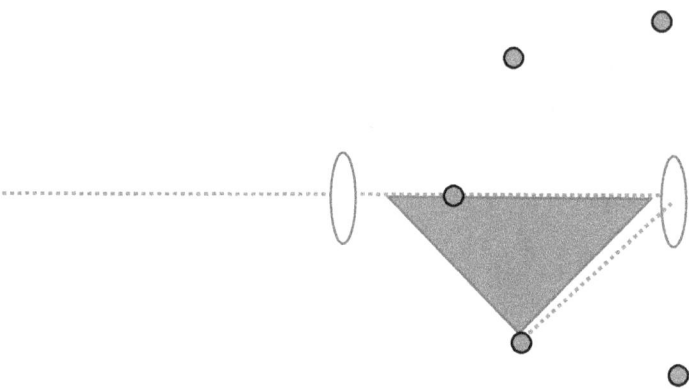

The above is the engineering of water. The nearest electrons are 104.5 degrees way in three directions creating a triangular valley where the Hydrogen electron wants to settle. The Hydrogen electron sits well within the fulcrum distance. H2O has a very strong.

Bonding 3D from Back of Hydrogen View

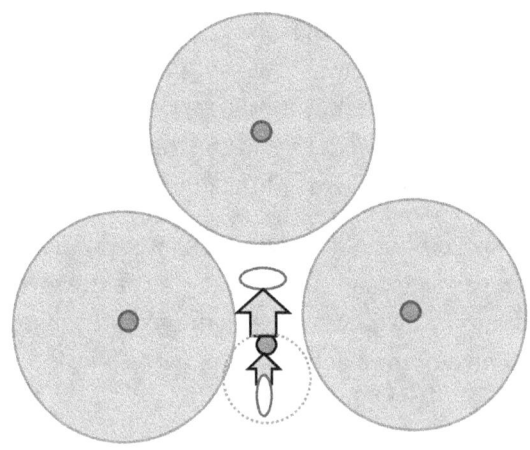

The basic concept of gas is that once an atom or molecules has enough heat, they are rotating such that not temporary bonds (liquid) occur. For all electrons outward molecules that calculation is the same. They will separate to the same ratio (effectively the formula is the extra energy of spacing molecules closer in a closed space versus the gravity constant of atoms)

The window closes by the molecule rotation before any bonds can occur. That is the gas state.

As the molecule rotates (by heat), one outer electron takes a path (a structure rotation, not a gravitation 'Bohr-Rutherford orbit'). That means that one outer electron obscures in 3D a whole circle of potential temporary (liquid) bonding positions. Over time, the nucleus is 'shaded' for an entire range of inclinations/longitudes.

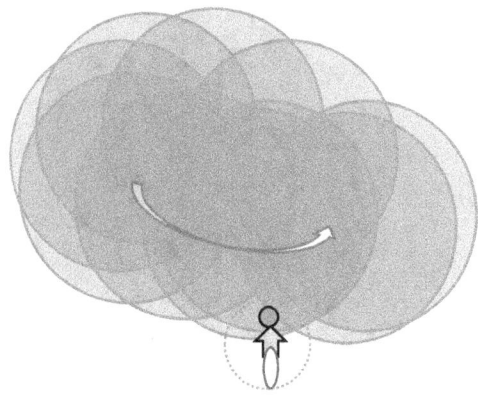

With all the electrons, the entire sphere for bonding is obscured, and the gas state begins.

However, for H2O, the calculation is different. It does not follow the Closed Space Ideal Gas Law. H2O has that Proton outward.

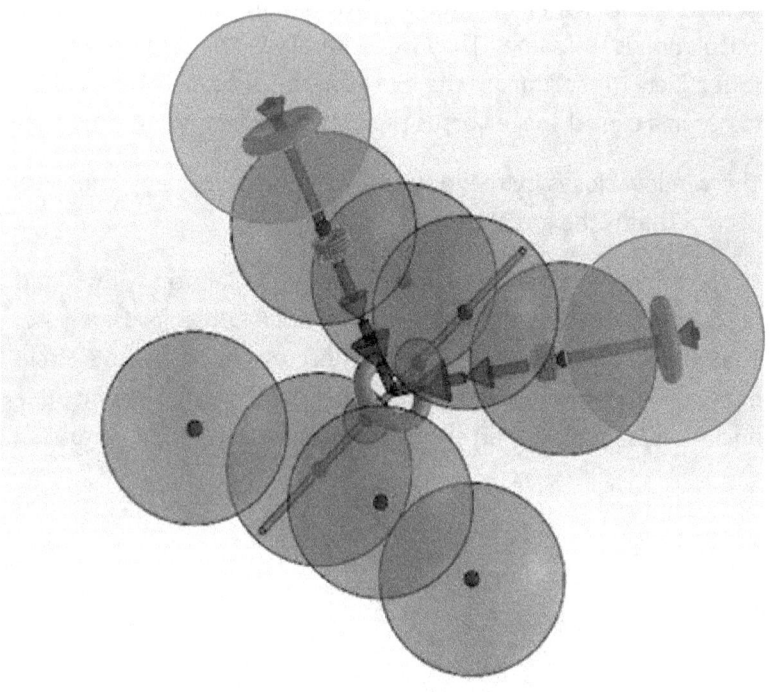

You must spin very fast for other electrons to obscure the outward Hydrogen nucleus proton form bonding.

That is why H2O has a very high boiling point relative to other molecules without Hydrogen proton outward.

Water and pH

Normal water has both Oxygen electron outward (6) and proton outward (2 Hydrogen). That ratio is 3:1, but the Hydrogen stick out, so in reaction the ratio is closer to 1:1 in reactivity (0.42:0.58) with the proton slightly more powerful.

Type	Direct Ratio	Strength	Reaction Ratio
Electron outward	6	1	6 (0.42)
Proton Outward	2	4 (2^2)	8 (0.52)
Total	8		14 (1.00)

The actual calculation is more complex with statistics which I leave to endnotes.[iv]

Sizing Hydrogen Proton Attraction Zones as Larger than Electron Repulsion Zones

In that sense, the prior drawing was not size correctly. Because the Hydrogen protons stick out, they are more reactive. In the engineering sense, the strength or the attraction field is about two times larger.

That means in AVSC, when we build Attraction Zones, they must be sized relative to the proton position versus the other atom's outer electrons. The more it sticks out, the relative large the Attraction Zone.

H2O Attraction and Repulsion Zones

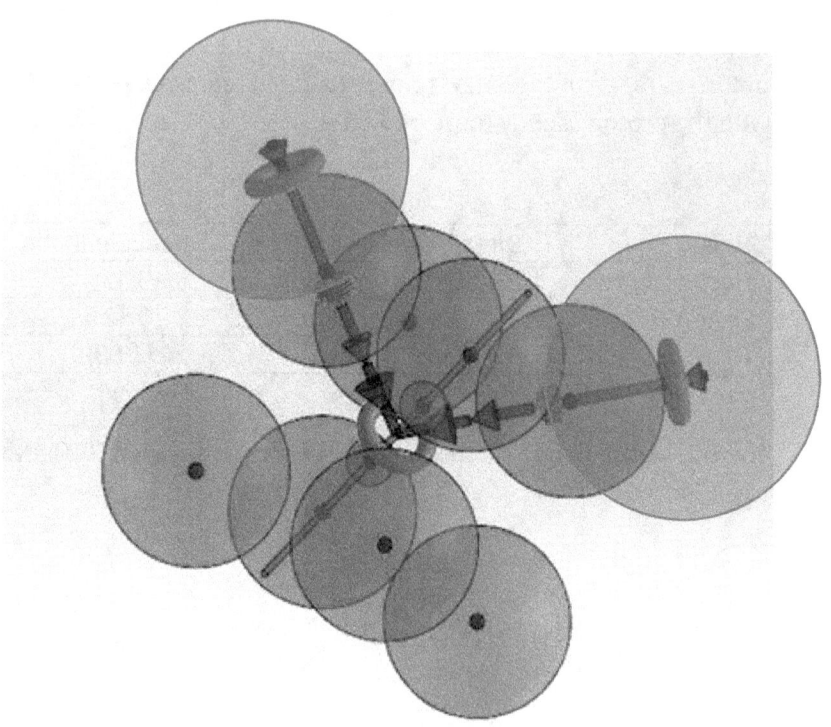

Percentage Reactivity of Hydrogen (pH)

In AVSC, the free protons (H+ Hydrogen) act like electrons do in ion. They fill up empty position. The water becomes a new semi-permanently bonded super-molecule.

Just like O-- ions are really with electrons that are filling the open positions, low pH Water takes up the extra protons to become a fully surrounded H2:6H+ relative stable molecule.

Low pH Means Water has Extra Hydrogen on Every Free Oxygen Electron

Low pH means that there is extra H+ Hydrogen nucleus, a proton only, in the water.

Moreover, in terms of the reaction process, this means that the Oxygen positions in H2O attract free H+ protons.

Moderate pH H2O Attraction and Repulsion Zones

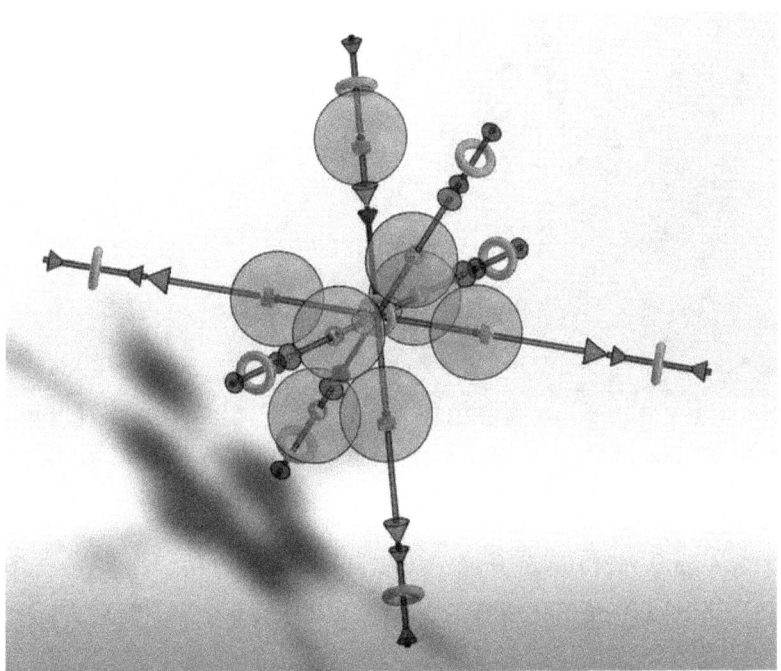

As a result, the entire molecule system them becomes all repulsion. This is the opposite environment of most molecules, all electrons outward, or even water which is effective half-attractive and half repulsive.

Low pH H2O Attraction and Repulsion Zones

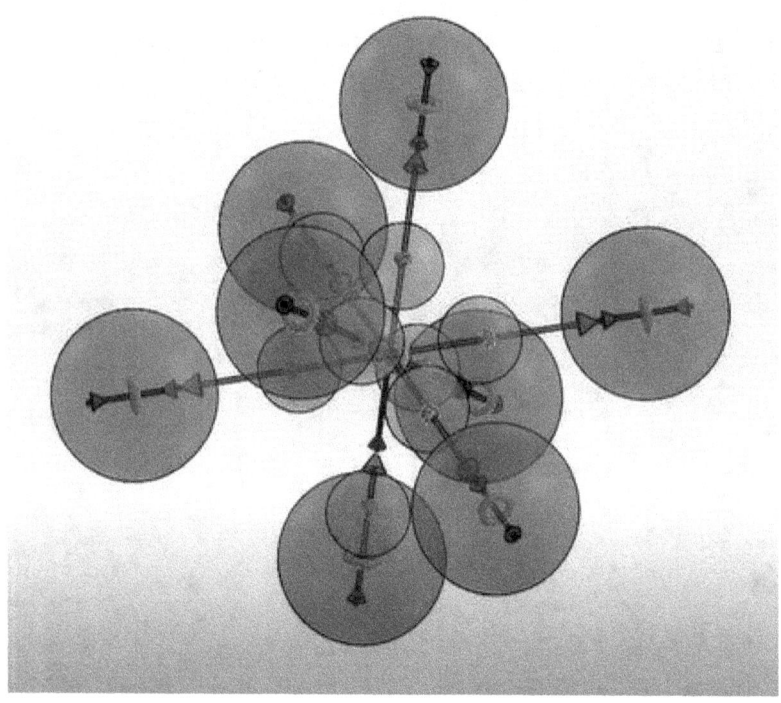

This means that the proton repulsion of the H2O:6H+ structure put direct pressure on any free H+ protons to bond. This changes the profile.

In fact, the deeper the tank, the more the pressure. This changes reaction speeds and causes reaction to occur.

The Thylakoid CO2 Breaking Photosynthesis Process

Much effort has been placed on reaction sites PH-1 and PH-II in research. Those assemblies exist on the Thylakoid cell wall. However, I have not found investigation into the whole Thylakoid itself as a reaction site. Mostly because it does

The purpose of the Thylakoid is to create a low-pH, proton rich environment. In that environment, reaction that would not normally occur will occur.

1) The gradient for H+ reactivity is hugely better than a standard 01-H atom.

The standard H plus has a Repulsion Zone that almost encompasses the proton-nucleus. The distance of the nucleus Receiver is less than the distance to the proton. That receiver is about 0.6×10^{-6}m.

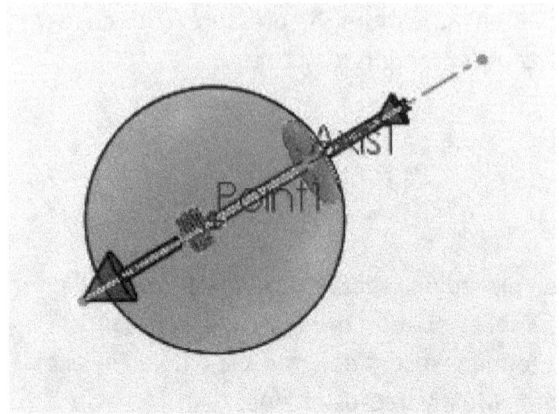

When H+, the proton has not offset for the electron. That makes the receiver about 2.5×10^{-6}m. Much longer than a standard Hydrogen atom.

However, when the proton is alone, the Receiver assembly:

- Is on bond sides creating bonding opportunities in every direction

- Is 4x longer at the same strength?

Standard Hydrogen Atom does not Break Bond with CO2

But when the repulsion zones are larger than the receiver bonding distance, a bond does not occur. This is the case for a full Hydrogen and CO2.

In the CO2 to full Hydrogen atom below, the repulsion zone of the Hydrogen electron extends past the proton-nucleus. As such, the two repulsion zones will repel before the CO2 electron can create a new bond to replace the Carbon-Hydrogen bond.

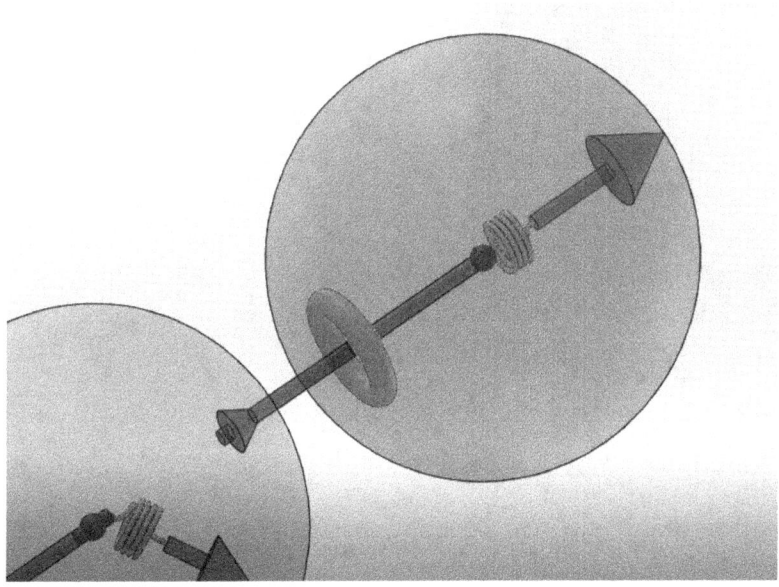

However, when the Hydrogen is in cation state (no electron), the receiving bonder is much longer. As such, it can create a bond to replace the Carbon-Oxygen weak double bond in CO2, and it will then create a new OH-structure.

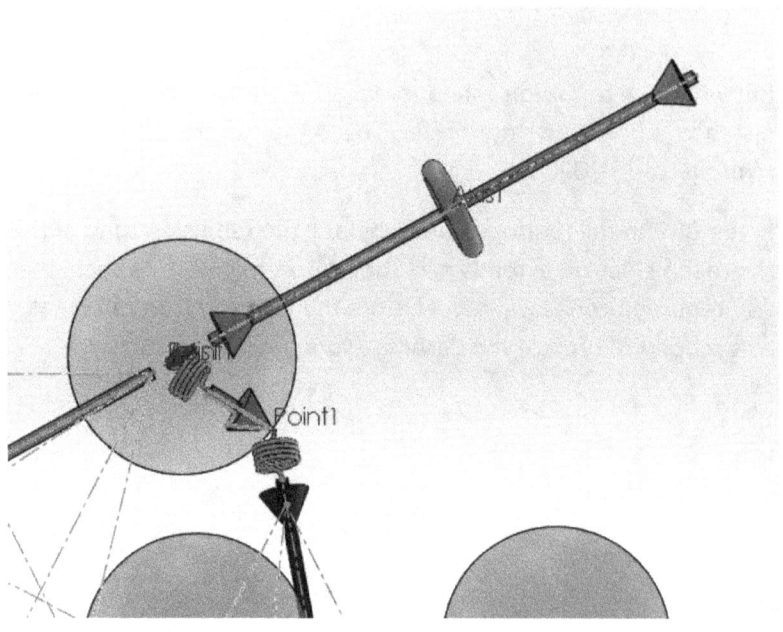

The electron shifts from the Carbon at the left to the Hydrogen. The connection to the Oxygen remains, and a new structure of O=C-OH (carbolic acid).

The initial structure of CO2 is O=C=O. A linear molecule with the C=O double bonds at opposite ends of the Carbon, and latitude offset by 90 degrees.

That new structure has different physical dimensions. The OH rotates away from the carbon. One O=C remains unchanged. That leaves

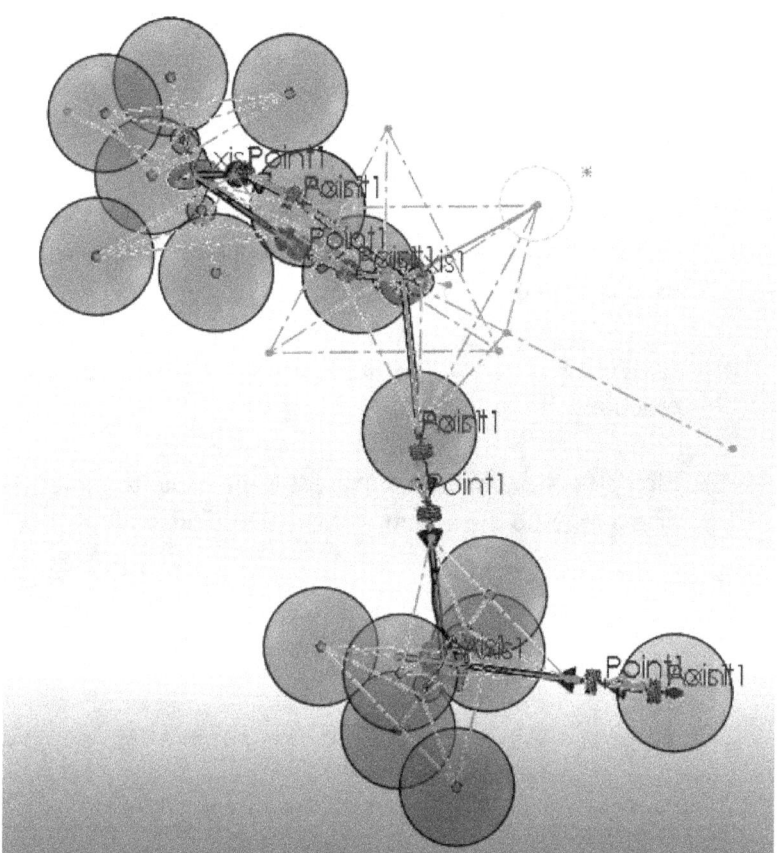

That structure still has an open position for an electron or another atom off the Carbon (yellow circle). In photosynthesis, this is passed to another assembly that adds that in the ADP <> ATP cycle.

So,

2) A standard pH-water environment would absorb those extra H+ reactants as semi-permanent bonds with the H2O exterior Oxygen electrons. Water at 5 pH would have H3O+ molecules as well as H2O molecules. There is

little available in the environment for H+ reactions. That is, even if H+ that H has a strong reaction, they are more closely bonded with H2O and so profile as if a standard H, with lower reactivity.

3) The low pH-water environment has free H+ protons which react with CO2 to create COOH, carbolic acid which is then used in the Kelvin Cycle.

4) Light is not required for the Thylakoid CO2 bond-breaking reactions.

5) High pH-water environment creates molecule-to-molecule pressure to up the endothermic reaction possibilities.

Hydrogen Bonds and Water

Remember how Hydrogen has 1) a Contributing bonding position from the electron and 2) a Receiving bonding position at the Nucleus Proton.

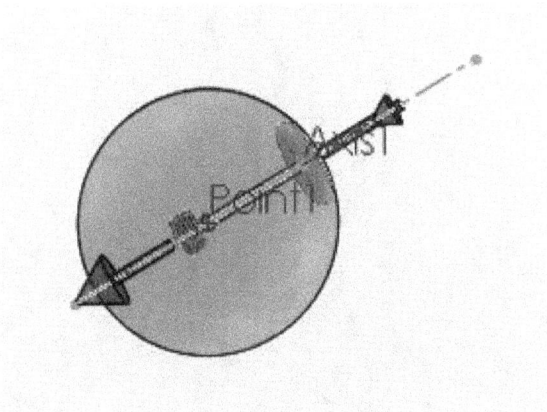

That means that Hydrogen bonds with anything, but just as importantly, after it bonds, Hydrogen effectively replaces the open position with one further out. If you have a receiver, add the Hydrogen, and you still have a receiver.

    a) A contributing Hydrogen contributes an electron inward toward the other atom, but the Hydrogen nucleus is outward as receiving, then seeking another atom's contributor.

b) A receiving Hydrogen receives an electron inward, but the Hydrogen electron is outward as receiving. This manner the combined molecule.

In AVSC Chemical Engineering Tool, Electrons (and Protons) have Repulsion Zones

Generally, an atom is a nucleus with surrounding electrons. Electrostatic force drives repulsion between the outer electrons; that is what makes atoms tend to remain the same element, and not change all the time.

Attached to every AVSC electron is a Repulsion Zone that provides an engineered method to make particles, of the same electrostatic charge, including electrons vs electrons, stay separated:

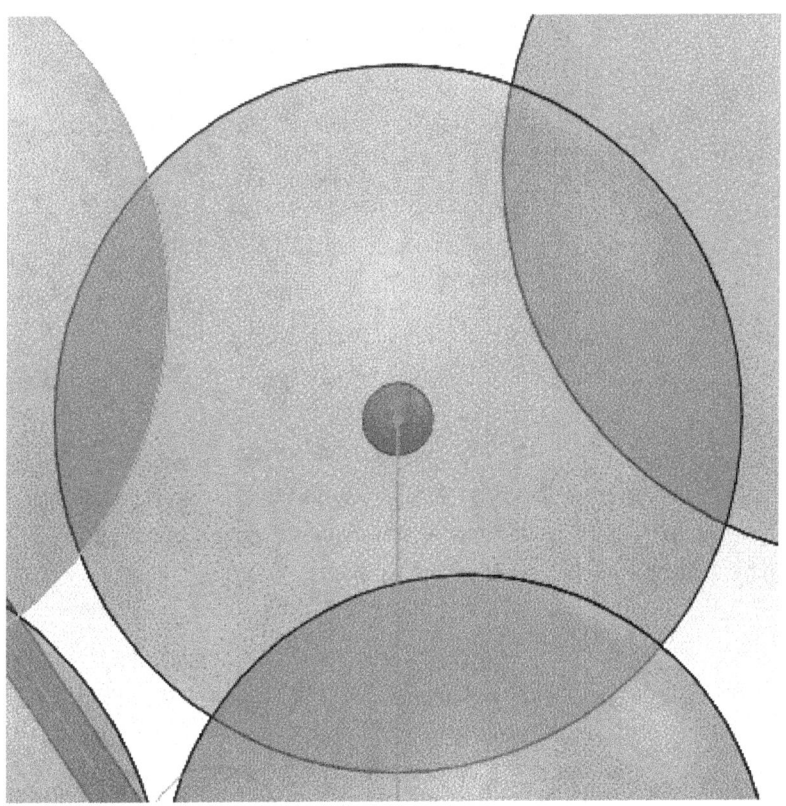

Repulsion Zones are engineered such that when the repulsion zones of two electrons meet they do not penetrate, but instead bounce off each other. In our preferred use, the material is translucent plastic which deforms slightly, and gives that acceleration back to the touching set of particles/fields.

When they meet in AVSC Animations, they bounce off each other.

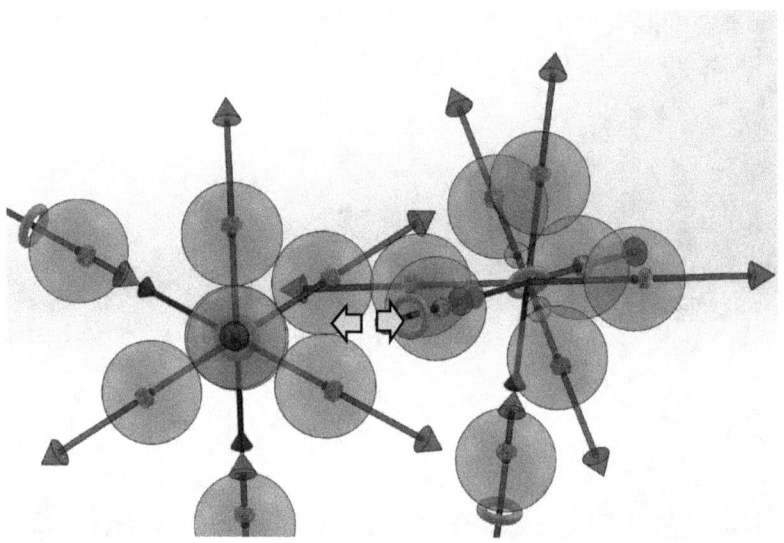

This common reaction mimics the way in which molecules, with exterior electrons tend to interaction - - - by bouncing off each other, keeping the atoms intact as a set.

A bond needs an open channel with extra spacing for other electrons that give them a connector long enough to reach into the receiving atom without bumping into other repulsion zones.

In AVSC Engineering, we have separated the distance of covalent bonding between three basic segments. Atom #1 has a Receiver, Atom #2, has a distance from its nucleus to the outer electron which will form the bond, and between those two is a Contributor attached to the outer #2 electron.

Those distances of the three components are not equal. In fact, the most interesting aspect in 3D is that bonding electrons with their contributor must find a recessed receiver. If the outer repulsion zones are too close, or in blocking alignment with other electrons, then the bond is unlikely.

This is why many molecules and reactions that one might think possible or likely do not occur. In AVSC, we can engineer that failure in 3D graphics. You can see that atoms cannot fit together even if there is bonding position. (I will describe the 3D nature of AT bonds and GC bonds in DNA strand in a later example.)

In a simple example, a connector from Atom #2 must find a receiver in Atom #1 without Repulsion Zone blockage. This is done with the three segments connecting two atoms.

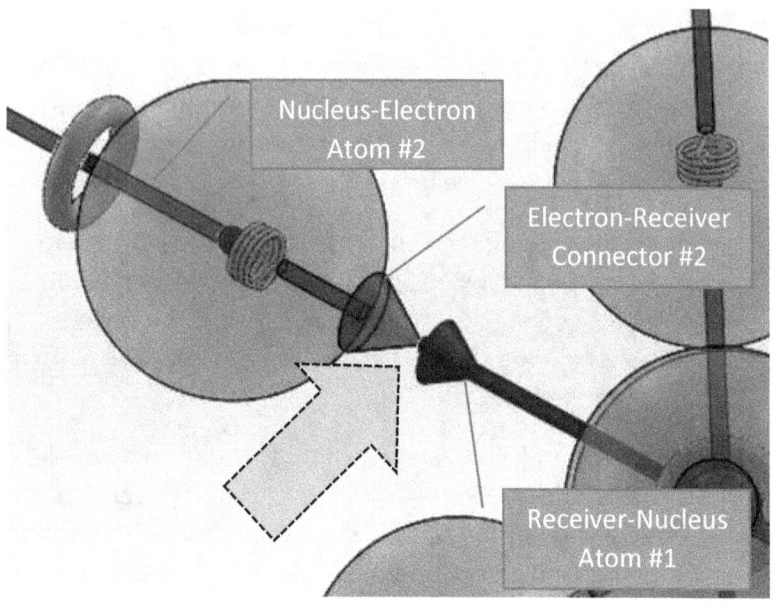

When Atoms Meet, the Repulsion Zones Operate Like Rubber Balls

On the other hand, sometimes the receiver does not find a contributor, and in that way, when repulsion zones meet, the atoms repel each other. In AVSC, repulsion zones are translucent rubbery balls around every electron particle.

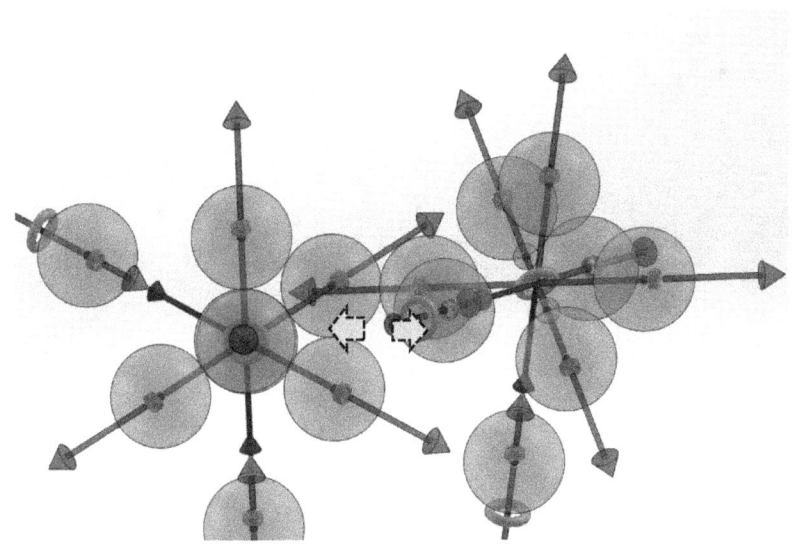

Equator View: Outer Subshell Structure
07-O Oxygen

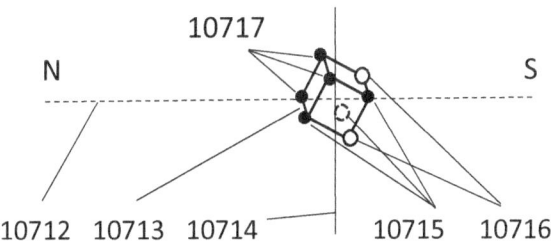

10717

N   S

10712 10713 10714   10715 10716

Understanding Bonding Angles for Carbon (109 degrees), Nitrogen (107), and Oxygen (104)

Hydrogen Drives Exceptions to the Closed Space Ideal Gas Law

Almost all atoms follow the Close Space Ideal Gas Law:

$$PV = nRT$$

P = Pressure

V = Volume

n = number of molecules

R = Constant

T = Temperature

That makes pressure go up linearly with Temperature, and temperature decrease as the volume is increase.

In that, once rotating, the extra negative electrostatic change in every direction becomes ubiquitous. In AVSC, every molecule is a structure with positive nucleus with electrons at the same average distance from the nucleus. That makes the calculation this linear relation, and all molecules give the same spacing in 3D.

In that way, the exteriors of atoms never form bonds, so the distance becomes dependent on the closed container volume, and you can calculate the pressure, volume, and temperature in ratio to the number of gaseous molecules.

The general rules are:

- Exterior of atoms are all electrons

- All atoms in gas state rotate fast enough that no bonds can occur (open paths close faster than any other molecule can approach to bond)

As a result, all gaseous electron-outward molecules generate the same surrounding space (at the same temperature, pressure, of course). A Neon is not different than a Di-Oxygen molecule.

However, that calculation is changed if the entire exterior is not electrons. If there is a proton-exterior, then the atom may never bond (be a gas), but the force of the electrons in the exterior will get reduced by the percentage of the sphere which is positive charged instead.

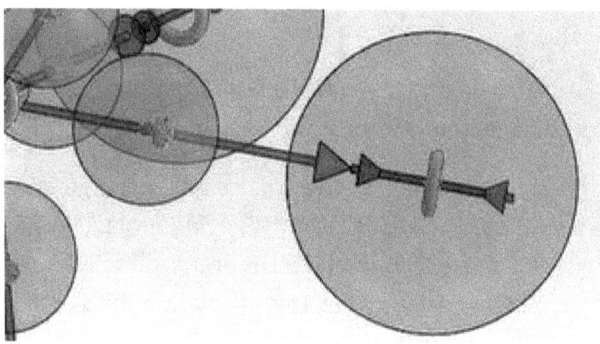

Conclusion

The great advance to science proposed here is that for many Chemical Engineering problems AVSC get calculations back towards physical size, position, and speed. This model updates and replaces fully the current angular momentum model by adding, to the AVSC expanded-classical model, an electron-nucleon nucleomagnetics repulsion force to create a complete equilibrium theory. That provides the knowable forces, separate electrostatic force and the postulated nucleomagnetics (magnetic fields at the particle level) that make the electrons move in unusual directions, specifically, then known energy level (position) which create the AVSC Periodic Chart of Elements which describes the attributes, like electrical conductivity, bonding, and such of each Element.

Please keep an open mind that the Arno Vigen Scrunched Cube (AVSC) nucleomagnetics expanded classical model delivers a certain range of engineering solutions, even if your university had drilled into you all the discontinuities of time and space as the reality. Be patient. We will argue for a few decades about my updated theoretical model, but in the meantime, what us engineers need, and can use, is functional tools at atomic and subatomic distances:

- to engineer;
- to visualize atoms, molecules and bonds;
- to create animations of interactions at the particle level, particularly chemical reactions that can occur or not occur based upon particle level geometry, not just tables of relative electronegativity or statistics; and
- to calculate size, position, velocity, force, and their direction for subatomic particles within reasonable tolerances.

The Arno Vigen Scrunched Cube (AVSC) Atomic Model adds a nucleomagnetics force for every particle, and that drives the first two electrons in every shell into closer positions at the poles. Hence, we get 01-H Hydrogen and 02-He Helium only in the first Shell, and then we get two electrons (subshell-s) in every shell at tight positions (higher release energy) – that is, scrunched.

As I said, a picture is much easier for engineer types. For an atom of 08-N Neon, two electrons (2m2) take the Shell-1 magnetic axis positions in blue. The either remaining electrons create a cube, with the two electrons at the nucleomagnetics axis 'scrunched' because the magnetics field strength is different there.

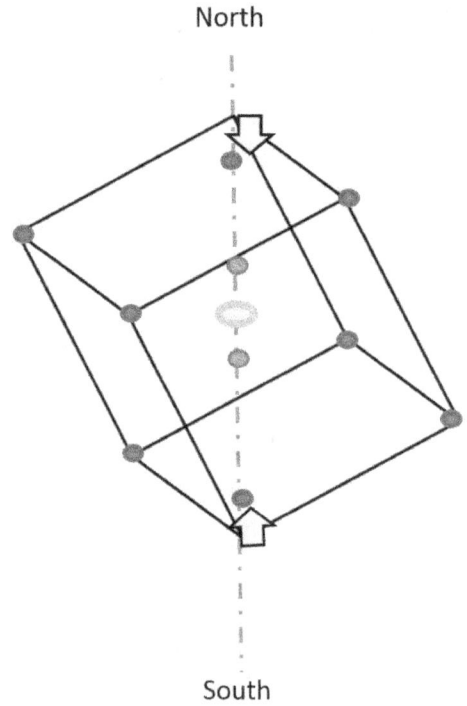

That said, I like build atoms, and molecules, and discovering in 3D how chemical reactions occur. These tools can do it better than statistical models – even if parts of the underlying theory get revises in the decades to come.

There are millions of derivative engineering work with the easy tools of AVSC. With these tools, chemistry is now a workable structure that can use generally traditional engineering CAD exactly when the simple solutions cannot provide answers. This presentation is just a tiny part of endless potential. I am sure you have ideas and questions. Please join me in a great adventure to advance science understanding and application.

Big hugs, let's get started.

Arno

Arno Vigen Scrunched Cube (AVSC) - Period Table of Elements

Shell Nbr	Group P01	Group P02	Group T03	Group T04	Group T05	Group T06	Group T07	Group T08	Group E01	Group E02	Group E03	Group 12	Group 13	Group 14	Group 15	Group 16	Group Minus-1	Group Inert
	Mag. Polar	Mag. Polar				Endcap Motomagnetic				Equatorial-90 High Electrical Conductivity		Metals		Scrunched Cube			Halogens	Inert Gases
1	001-H Hydrogen 1.0																	002-He Helium 4.00
2	Single / Double								003-Li Lithium 6.9	004-Be Beryllium 9.0	005-B Boron 10.8	Scrunched Cube Corners	006-C Carbon 12.0	007-N Nitrogen 14.0	008-O Oxygen 16.0	009-F Florine 19.0		010-Ne Neon 20.2
3									011-Na Sodium 23.0	012-Mg Magnesium 24.3	013-Al Aluminum 27.0	Faces of the Scrunched Cube	014-Si Silicon 28.1	015-P Phosphorus 31.0	016-S Sulfur 32.1	017-Cl Chlorine 35.5		018-Ar Argon 40.0
4	019-K Potassium 39.1	020-Ca Calcium 40.1	021-Sc Scandium 45.0	022-Ti Titanium 47.9	023-V Vanadium 51.0	024-Cr Chromium 52.0	025-Mn Mangese 54.9	026-Fe Iron 55.8	027-Co Cobalt 58.9	028-Ni Nickel 58.7	029-Cu Copper 63.5	030-Zn Zinc 65.4	031-Ga Gallium 69.7	032-Ge Germanium 72.6	033-As Arsenic 74.9	034-Se Selenium 79.0	035-Br Bromine 79.9	036-Kr Krypton 83.8
5	037-Rb Rubidium 85.5	038-Sr Strontium 87.6	039-Y Yttrium 88.9	040-Zr Zirconium 91.2	041-Nb Niobium 92.9	042-Mo Molybdenum 96.0	043-Tc Technetium 98.0	044-Ru Ruthenium 101.1	045-Rh Rhodium 102.9	046-Pd Palladium 106.4	047-Ag Silver 107.8	048-Cd Cadmium 112.4	049-In Indium 114.8	050-Sn Tin 118.7	051-Sb Antimony 121.8	052-Te Tellurium 127.6	053-I Iodine 129.9	054-Xe Xenon 131.3
6	055-Cs Cesium 132.9	056-Ba Barium 137.3	057-La>071 Series	072-Hf Hafnium 178.5	073-Ta Tantalum 180.9	074-W Tungsten 183.8	075-Re Rhenium 186.2	076-Os Osmium 190.2	077-Ir Iridium 192.2	078-Pt Platinum 195.1	079-Au Gold 197.0	080-Hg Mercury 200.6	081-Tl Thallium 204.4	082-Pb Lead 207.2	083-Bi Bismuth 209.0	084-Po Polonium 209.0	085-At Astatine 210.0	086-Rn Radon 222.0
7	087-Fr Francium 223.0	088-Ra Radium 226.0	089-Ac>103 Series	104-Rf Rutherfordium 267.0	105-Db Dubnium 268.0	106-Sg Seaborgium 269.0	107-Bh Bohrium 270.0	108-Hs Hassium 277.0	109-Mt Meitnerium 278.0	110-Ds Darmstadium 281.0	111-Rg Roentgenium 282.0	112-Cn Copernicium 285.0	113-Nh Nihonium 286.0	114-Fl Flerovium 289.0	115-Mc Moscovium 290.0	116-Lv Livermorium 293.0	117-Ts Tennessine 294.0	118-Og Oganesson 294.0

057-La Lanthanum 138.9	058-Ce Cerium 140.1	059-Pr Praseodymium 140.9	060-Nd Neodymium 144.2	061-Pm Promethium 145.0	062-Sm Samarium 150.3	063-Eu Europium 152.0	064-Gd Gadolinium 157.3	065-Tb Terbium 158.9	066-Dy Dysprosium 162.5	067-Ho Holmium 164.9	068-Er Erbium 167.3	069-Tm Thulium 168.9	070-Yb Ytterbium 173.0	071-Lu Lutetium 175.0
089-Ac Actinium 227.0	090-Th Thorium 232.0	091-Pa Protactinium 231.0	092-U Uranium 238.0	093-Np Neptunium 2,337.0	094-Pu Plutonium 244.0	095-Am Americium 243.0	096-Cm Curium 247.0	097-Bk Berkelium 247.0	098-Cf Californium 251.0	099-Es Einsteinium 252.0	100-Fm Fermium 257.0	101-Md Mendelevium 258.0	102-No Nobelium 259.0	103-Lr Lawrencium 266.0

Endnotes

[i] https://upload.wikimedia.org/wikipedia/commons/e/e7/Hydrogen_Density_Plots.png

[ii] That means in quantum mechanics at energy levels below the Planck-Einstein quantum of energy, at real levels below Heisenberg uncertainty level.

[iii] AVSC = Arno Vigen Scrunched Cube Atomic Model as described in various engineering and education patents pending for chemical reactions including 15/490,870

[iv] The volume of approaches directly toward the nucleus is the straight 2 of 8 ratio for protons to electrons, yet from a tangential approach, even if it misses the electron, another atom might interact with the proton. Further, since the other electron may be all electrons outward, it only reacts with the proton, and
As such, the H2O calculation must be of
Further, the calculation is more complex because what really happens is that for the temporary bonds or permanent bonds, the electron of one H2O Oxygen atom aligns to the nucleomagnetics axis of the Hydrogen proton. That is, at the scrunched cube 66 degrees or nucleomagnetics axis.
You can read great references to this at University of Iowa's Chemistry Dept., and others that determined that water keeps bonding at this scrunched cube positions.
That means that once you get two bonds, there is the space from the other atoms such that the other Oxygen bonding positions do not tend to bond permanently. That is the structure of ice crystals.

www.ingramcontent.com/pod-product-compliance
Lightning Source LLC
Chambersburg PA
CBHW030058230526
45471CB00003B/1144